This Book Belongs To:

Made in the USA
Monee, IL
13 January 2022

88847840R00069